CATERPILLAR THIRTY
PHOTO ARCHIVE

CATERPILLAR THIRTY
PHOTO ARCHIVE

Photographs from the
Caterpillar Inc. Corporate Archives

Edited with introduction by
P. A. Letourneau

Iconografix
Photo Archive Series

Iconografix
P.O. Box 18433
Minneapolis, Minnesota 55418 USA

Text Copyright © 1993 by Iconografix

The photographs and illustrations are property of Caterpillar Inc. and are reproduced with their permission.

All rights reserved. No part of this work may be reproduced or used in any form by any means--graphic, electronic, or mechanical, including photocopying, recording, taping, or any other information storage and retrieval system--without written permission of the publisher.

Library of Congress Card Number 93-78196

ISBN 1-882256-04-2

93 94 95 96 97 98 99 5 4 3 2 1

Cover and book design by Lou Gordon

Printed in the United States of America

PREFACE

The histories of machines and mechanical gadgets are contained in the books, journals, correspondence and personal papers stored in libraries and archives throughout the world. Written in tens of languages, covering thousands of subjects, the stories are recorded in millions of words.

Words are powerful. Yet, the impact of a single image, a photograph or an illustration, often relates more than dozens of pages of text. Fortunately, many of the libraries and archives that house the words also preserve the images.

In the Photo Archive Series, Iconografix reproduces photographs and illustrations selected from public and private collections. The images are chosen to tell a story...to capture the character of their subject. Reproduced as found, they are accompanied by the captions made available by the archive.

The Iconografix Photo Archive Series is dedicated to young and old alike, the enthusiast, the collector and anyone who, like us, is fascinated by "things" mechanical.

ACKNOWLEDGMENTS

The photographs which appear in this book were made available by the Caterpillar Inc. Corporate Archives. We are most grateful to Caterpillar Inc., and sincerely appreciate the cooperation of Joyce Luster, Corporate Archivist.

We also wish to thank Ziegler Inc., Minnesota-based Caterpillar dealer, for its loan of supplemental photographs.

Early Caterpillar Thirty with rear mounted seat.

INTRODUCTION

In April 1925, Holt Manufacturing Company and C.L. Best Tractor Company merged to create Caterpillar Tractor Company. Each company brought an existing line of track-laying tractors to the union. Holt, who had built its first such machine in 1904, contributed the 2 Ton, 5 Ton, and 10 Ton. Best, who had built its first such unit in 1913, contributed the Thirty and Sixty.

Best had introduced the Thirty in 1921. The company rated its performance at 30 belt and 18 drawbar horsepower. The Thirty was fitted with a vertical 4-cylinder gasoline engine, manufactured by Best, that operated at 800 rpm. It featured 4.75 x 6.5 inch bore and stroke. The tractor's 2-speed transmission offered operating speeds of 2 and 3.0625 mph. Maximum drawbar pull was rated at 4,343 lbs. By 1923, the Thirty's performance was improved, and the tractor was re-rated at 30 belt and 20 drawbar horsepower. Forward speeds were increased slightly to 2.03 and 3.1 mph. Maximum drawbar pull was rated at 4,930 lbs. In 1924, Best again upgraded the Thirty's performance. Its engine was fitted with a new carburetor, and operating speed was increased to 850 rpm. The tractor was equipped with a new 3-speed transmission, that offered operating speeds of 1.75, 2.63, and 3.63 mph. Thus equipped, the Thirty offered a maximum 37.83 brake horsepower and 7,563 lbs drawbar pull.

The Thirty's greatest strength was its ability to operate effectively in almost any soil condition. It offered greater traction and flotation than were available with wheel tractors. While the Thirty was more commonly used in agriculture than was the larger Sixty, it was also popular in logging, road building, and construction operations.

The tractor changed little after the formation of Caterpillar, although a low-compression engine that permitted the burning of kerosene was eventually offered. The Thirty remained in production until 1931. It was manufactured both at San Leandro, California (S Series) and Peoria, Illinois (PS Series). In total, and including Best production, 9,536 units were built at San Leandro; 14,292 units were built at Peoria. Enthusiasts should note that the Caterpillar Inc. Corporate Archives include dozens of photographs of the Thirty, but very few from the years before the Holt and Best merger. For this reason, only one photograph of the Best Thirty is included in this book.

Five years after the S and PS Series Thirtys were retired from production, Caterpillar introduced a new tractor that also carried the designation Thirty. This tractor is sometimes labeled the 6G Series Thirty, as "6G" was the prefix to its serial number sequence. In 1937, after Caterpillar had built 875 units, the tractor was redesignated the Caterpillar R 4. It remained in production until 1944. We include photographs of the later series Thirty in this book, with the full knowledge that it had little in common with the earlier series.

The 6G Series Thirty was available with either a low-compression, distillate burning or high-compression, gasoline burning 4-cylinder engine, that operated at 1,400 rpm. Its bore and stroke measured 4.25 x 5.50 inches. Its 5-speed transmission offered operating speeds ranging between 1.7 and 5.4 mph. The low-compression, distillate version developed a maximum 34.13 brake and 26.71 drawbar horsepower; and a maximum 6,120 lbs drawbar pull. The high-compression, gasoline version developed a maximum 37.81 brake and 30.99 drawbar horsepower; and a maximum 7,211 lbs drawbar pull.

S and PS Series Thirty

Thirty with top mounted seat introduced in 1926.

Caterpillar Thirty equipped for winter work. December 1929.

Thirty and rotary scraper working on highway. Bad Lands National Park, South Dakota. August 1931.

Thirty with Willamette double drum winch operating a .75 yard Crescent drag line bucket. Crispus Burn section of Rainier National Forest. January 1929.

Caterpillar Thirty with Giant Premier No. 6 grader working on streets in a California oil field. May 1927.

Thirty and Ball scraper.

Thirty with Ball scraper at Redwood City, California.

Thirty pulling an Adams grader. Preparing shoulders for the Caterpillar Trail, a scenic highway along the Illinois River. August 1928.

Thirty with root plow. The second operation in preparing the subgrade on the Caterpillar Trail. Peoria, Illinois. August 1928.

Thirty with bulldozer working on U.S. Highway 66. Gallup, New Mexico. July 1931.

Thirty with 8 foot blade grader fine grading ahead of paver. June 1928.

Thirty operating three Euclid scrapers on state highway road work. Loudoun County, Virginia. April 1930.

Thirty with early type of elevating grader building roads near Sioux Falls, South Dakota. August 1929.

Ethiopian Emperor Heilie Selassie (bearded), accompanied by his sons, views Caterpillar Thirty and grader at work on the Jhimma Road. Wolisso, Ethiopia. November 1932.

Two Caterpillar Thirtys pulling a Sixty L.W. grader fitted with two 3 foot moldboard extensions. March 1931.

Thirty with three Baker Maney self-loading scrapers working on approach to bridge.

Thirty and Ball wagon grader building a road near Seville, Georgia. June 1931.

Thirty with Russell Special grader. May 1930.

Thirty and road maintainer. April 1926.

Thirty and Hi-Way Service scarifier ripping up old road, preparatory to resurfacing. August 1931.

Thirty equipped with beet shoes pulling Killefer beet plow. Rio Vista, California. September 1928.

Caterpillar Sidehill Thirty with Killefer double disk. Vacaville, California. February 1929.

Thirty pulling 3-bottom Oliver plow. Rio Vista, California. October 1926.

Disking with Caterpillar Thirty. August 1929.

Thirty disking in 1100 acre field. Liberal, Kansas.

Disking bull thistle with Caterpillar Thirty and Dinuba Steel Products disk. July 1928.

Caterpillar Thirty pulling a 5-disk, 26 inch sugar land plow.

Rear Seat Hill Special Caterpillar Thirty disking in orchard. Watsonville, California. April 1929.

Thirty furrowing for irrigation on a date ranch. Indio, California. September 1930.

Caterpillar Wide Gauge Thirty pulling four Athey wagon loads of sugar cane. Lake Okeechobee, Florida. February 1933.

Caterpillar Thirtys pulling 16 foot International Harvester combines in 1400 acre wheat field. Walla Walla, Washington. August 1928.

Eighteen Caterpillar Thirtys and combines working in wheat field. The owner of 40,000 acres operated 36 Caterpillar combines. Hays, Kansas. August 1931.

Best Thirty and Case combine. August 1922.

Thirty pulling two 10 ton wagons through soft fields on a Sherman County, Oregon wheat ranch. August 1930.

Thirty and Holt combine.

Thirty and Holt combine with side hill arrangement. Murdo, South Dakota. August 1928.

Thirty with two binders. Sioux Falls, South Dakota. October 1927.

Thirty with Holt combine. Faulkton, South Dakota. October 1928.

Caterpillar Thirty pulling Caterpillar No. 3 combine in Lubbock, Texas grain field. July 1931.

Thirty and Splittstoser 3-row potato digger. Moorhead, Minnesota. October 1930.

Thirty pulling three wagon loads of baled hay. Alviso, California. August 1928.

Rear seat and Holt Western combine.

59

Caterpillar Thirty pulling Ronning ensilage harvester. September 1929.

Caterpillar Thirty pulling 1.5 yard tumble bug scraper. Browns Valley, California.

Thirty logging in Oregon woods near Klamath Falls.

Caterpillar Thirty with high wheeler. Klamath Indian Reservation. Oregon. October 1928.

Caterpillar Thirty and Fifteen at the entrance of the Hollow Log Garage. Sequoia National Park. December 1931.

Thirty pulling 20 cords of pulp wood. Port Arthur, Ontario. 1929.

Thirty equipped with post hole digger. December 1929.

Thirty using a Willamette double drum winch to rig up an oil well. Texas. April 1929.

Thirty transporting 8,000 pound roll of cable up steep hill in Dublin Canyon, California. March 1929.

Thirty with winch stringing cable. California. October 1928.

Thirty with Killefer subsoiler pulling bare copper wires through the ground. Louisville, Kentucky. May 1930.

Thirty equipped with Highway trailer reel cart and winch. Ashland, Virginia. June 1930.

Wide Gauge Caterpillar Thirty equipped with Cardwell All-Steel side boom drawing nine strands of one inch copper cable.

Thirty switching boxcars loaded with empty tin cans. Chicago, Illinois. March 1929.

The first Caterpillar in Martinique, West Indies; captured on a glass plate negative.

Thirty with W-K-M side boom unloading 22 inch, 20 foot pipe. Stinnett, Texas. July 1930.

Thirty cold bending high carbon 12 inch electric welded oil pipe. Two Thirtys equipped with Allsteel side booms. East Texas. September 1931.

Thirty with Ateco hydraulic scraper working in sand dunes of San Francisco. April 1930.

Thirty with LaPlant-Choate hydraulic bulldozer. July 1927.

Thirty operating at Forest Lawn Memorial Cemetery. Glendale, California. April 1929.

Thirtys with wagons working under 1.25 yard gasoline shovel. One Thirty equipped with hand operated bulldozer, the other equipped with Willamette single drum winch. Toledo, Ohio. November 1927.

Two Caterpillar Thirtys and one Sixty at work grading Fort Wayne Municipal Airport. November 1928.

Thirty hauling coal at Chicago brickworks. July 1930.

Thirty equipped with bulldozer terracing near Washington Monument. January 1932.

Thirty equipped with W-K-M side boom and clamshell bucket digging foundation for steel utility towers. Indianapolis, Indiana. July 1930.

Steam shovel loading wagons pulled by a Caterpillar Thirty.

Thirty with Killefer loader operating in brickworks. Chicago, Illinois.

Caterpillar Thirty pulling rail car towards a B&O Railroad car repair shop. Garrett, Indiana.

Thirty with bulldozer used to smooth off coal after it was unburied. March 1931.

Thirty with Euclid wagon at work on the new stadium at the University of Virginia. Charlotteville. October 1930.

Early Caterpillar Thirty equipped with optional hood and engine side panels.

Thirty with extended track roller frame, 5-roller under carriage and street pads.

Thirty equipped with optional canopy and belt drive.

Thirty equipped with optional pintle hitch.

Late Thirty with rear seat.

Caterpillar Thirty with fully enclosed cab.

Caterpillar Thirty with factory cab. January 1929.

Caterpillar Thirty with optional bumper and canopy.

Thirty with improved heavy duty spring and front idler assemblies.

Thirty with Williamette single drum towing winch manufactured in Portland, Oregon.

Thirty with snow plow.

6G Series Thirty

Caterpillar Thirty. 1937.

Thirty Orchard Model with top seat.

Caterpillar Thirty engine.

Butane Model Thirty.

Butane Thirty (foreground) and Caterpillar RD-6.

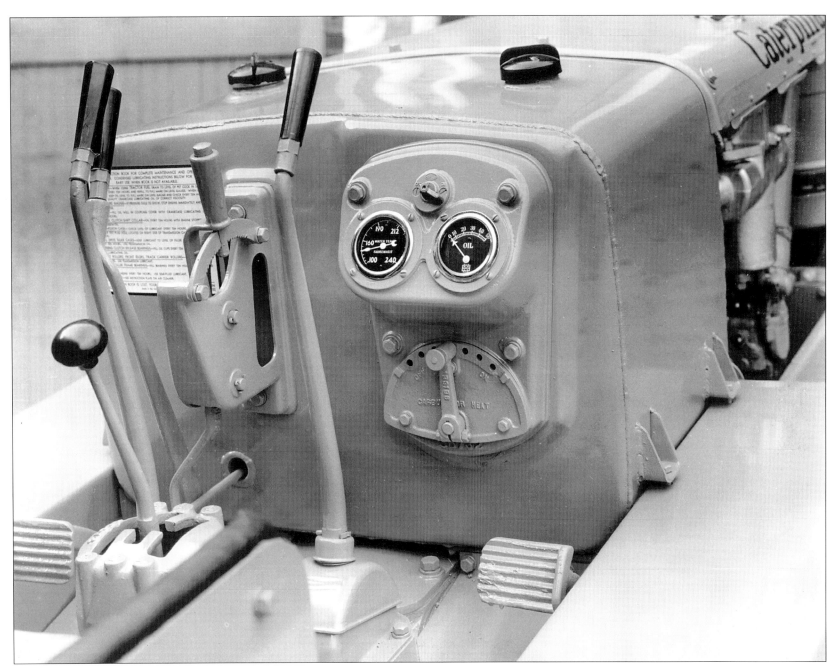
Thirty operator controls and instruments.

Thirty with Anthony Model B multiple tool shovel. June 1937.

Thirty with Trackson sideboom and winch moving tank sections for a new refinery. Lovell, Wyoming. August 1937.

Trackson high-shovel mounted on Caterpillar Thirty. Milwaukee, Wisconsin. 1937.

Excavating and loading with Thirty and Anthony loader. Peoria, Illinois. July 1937.

Thirty with LeTourneau bulldozer. Oklahoma City, Oklahoma. April 1937.

Wrecking building with Caterpillar Thirty equipped with 4-drum, power controlled unit and LeTourneau tractor crane. Shown lifting out steel girders. Peoria, Illinois. March 1936.

Thirty with Trackson Loader digging a basement. East Peoria, Illinois. November 1937.

Grading shoulders on a California state highway project with Caterpillar Thirty and No. 33 grader. March 1937.

Thirty and Killefer rotary scraper tearing down levees for the Leslie Salt Company. Newark, California. January 1938.

Thirty skidding logs near Lisbon, Ohio. May 1936.

Thirty pulling 8.5 foot Killefer chisel and 2-section harrow. Camerillo, California. April 1936.

Thirty pulling a 9.75 foot disk. Santa Maria, California. May 1936.

Thirty pulling John Deere 2-bottom, 2-way plow in alfalfa ground and gumbo soil. Fort Morgan, Colorado. April 1936.

Caterpillar Thirty and Caterpillar No. 2 terracer at work on the farm of Aetna Life Insurance Company. Lexington, Oklahoma. February 1936.

Thirty pulling 3-bottom plow in old alfalfa field. Oxnard, California. February 1936.

Thirty and sloper hop plow making 8 foot rows between hop vines. The men riding the plows guide outside cutters close to vines. Albany, Oregon. April 1937.

Thirty pulling a 12.5 foot disk and three 6 foot Meeker harrows. Bridgeton, New Jersey. July 1936.

Thirty and No. 22 terracer. Hickman County, Kentucky. April 1937.

Thirty pulling 20 foot Rumely combine. Stafford County, Kansas. July 1937.

Caterpillar Thirty pulling Model 36 combine.

Thirty pulling Killefer 8 foot disk in apricot orchard. San Martin, California. April 1936.

Thirty pulling 9.5 foot disk in pear orchard. Agnew, California. April 1936.

The Iconografix Photo Archive Series includes:

JOHN DEERE MODEL D Photo Archive   ISBN 1-882256-00-X
JOHN DEERE MODEL B Photo Archive   ISBN 1-882256-01-8
FARMALL F-SERIES Photo Archive   ISBN 1-882256-02-6
FARMALL MODEL H Photo Archive   ISBN 1-882256-03-4
CATERPILLAR THIRTY Photo Archive   ISBN 1-882256-04-2
CATERPILLAR SIXTY Photo Archive   ISBN 1-882256-05-0
TWIN CITY TRACTOR Photo Archive   ISBN 1-882256-06-9   *(Sept. 93)*
MINNEAPOLIS-MOLINE U SERIES Photo Archive   ISBN 1-882256-07-7   *(Sept. 93)*

The Iconografix Photo Archive Series is available from direct mail specialty book dealers and bookstores throughout the world, or can be ordered from the publisher.
For information write to:

Iconografix
P.O. Box 18433
Minneapolis, Minnesota 55418 USA